Explosion

Explosion

Some Sources, Causes, and Effects in Man's Affairs

Douglas W. Gould

VANTAGE PRESS
New York / Washington / Atlanta
Los Angeles / Chicago

FIRST EDITION

Published by Vantage Press, Inc.
516 West 34th Street, New York, New York 10001

Manufactured in the United States of America
ISBN: 0-533-06615-8

Library of Congress Catalog Card No.: 85-90101

Contents

Foreword

The dividing line separating us from disaster is seldom seen until after the brush with the adverse condition. I have had my share of "near misses" in war, peace, and industrial research; I have seen what happened on the other side of the dividing line. The events related here are recounted in detail in widely available sources; the conclusions are mine.

Introduction

EXPLOSION! Its newspaper headline can chill the blood. It is no common event. It is seldom foreseen as possible. There has been no feeling of brave resolution in the face of disaster. It is over before pen can be put to paper, let alone appearing in print. The survivors have had an unique experience. They have been exposed to an unknown, unseen force, and without reason or selection. No trace of the assailant is found, nor can the survivors describe the experience.

This is not a recital of disasters for the sake of shocking the reader, although it must include the circumstances attendant to the disaster. The criteria for inclusion center about immensity and the resultant impact on human thought. For large events bring many ideas together as to causes—a major disaster can prompt suggestions of means to prevent a repetition. The forty thousand fatalities attributed yearly to auto traffic in the United States seem to have little impact; an annual average of two hundred fatalities in civilian aviation is given little thought. But if there is a major commercial air accident involving fifty persons there is a tide of activity wherein airlines, plane manufacturers, and civil authorities struggle to assess causes, to allocate responsibility, and to propose means to avert a recurrence.

Some people would prefer that explosions pass unrecorded, or at least not publicized, particularly if the events are disasters. These people stand in contrast to those who would report, as noteworthy, *only* disasters. We shall be concerned only with explosions that are immense. It is impossible to weigh them equally in terms of explosive involved or lives lost or property damaged. In some instances the amount of explosive was beyond calculation, but no lives were lost and the property damage was moderate. But in our recital there must be one of these elements involved, and the choice inevitably includes the premise that Man's attention has been so gripped that he has been led to seek to understand the attending conditions, to find causes, applications, or some means of prevention. Understanding is the key to significance.

Each catastrophic disaster is followed by a wave of investigatory activity that seems to seek, equally, for "cause" and "blame"; and usually finds no surviving evidence. The tendency is then to draw a curtain over the event and, if possible, to drop it into a category somewhat like "acts of God." But other earnest persons find analogies and parallels between small accidents or near-accidents and disasters, and in this search is found the great mass of evidence that says that conditions of most explosions are avertible. The smaller the explosion, the greater the chance for survival of credible evidence. Improper environment, defective equipment, unauthorized acts—all these may give a picture of the elements that fortuitously coincide as the prelude to disaster. Simply stated, they give an unqualified statement that disaster is not inevitable; it is avertible. The keys lie in proper design, adequate inspection, intelligent discipline. Each of these factors partakes of the human spirit and knowledge and, if it were imposed on the ordinary explosion, would remove the unresolved aspects so often attendant to disaster.

Explosion

1. The Anatomy of an Explosion

A. General

We sometimes become involved when trying to describe an explosion; we may use cause and effect interchangeably. Technically, an explosion is the tremendous release of energy in a very short interval of time. What preceded or initiated the release is often ascribed to some unrelated factors. For our purposes, we may think of an explosion as the creation of a mass of rapidly expanding gas. The means of creating the gas or of causing it to expand are generally well understood. An explosion has the attribute of instantness. The term implies some lapse of time too brief to be perceived by and reacted to by the human senses. That this can be so is seen by the fact that impulses in the human nervous system move at the rate of not more than three hundred feet per second, while in an explosion, the shock wave may travel at several thousand feet per second.

An explosion is distinct from those other categories of human disaster or natural phenomena in which the unleashing of forces can visibly change the environment. Let us examine some of these other conditions. A river in flood (unless the flooding is due to the failure of a dam) is observable as it approaches a crest, and some measure of relief or escape from flood waters may be possible. After the flood has passed, one can search the mud for possessions, and some may be recovered downstream. Even in the event of the dreaded tidal wave *(tsunami)*, with today's communication, there is a chance that observation at a distant points may allow time for issuing warnings. Or consider a conflagration: often its onset may be seen; heroic measures are made to contain it. Even a holocaust like Chicago's of 1871 was attended by relatively small loss of life, despite the destruction of over two thousand acres of buildings. The fire spread slowly; citizens saw approaching danger and had time to retreat. A hurricane usually heralds its approach and, before building to maximum force, may allow escape laterally from its path or retirement to a cellar or resistant structure.

No such advantages are present at an explosion. No forewarning is given; no time for counter-measures; no instant to think. The blast is accompanied by no ameliorations. Usually those present at the point of the explosion are annihilated; this prevents a firsthand report as to why there was an explosion. Those who were also nearby but who escaped death are rendered incapable of lucid thinking. The conditions readily explain this. The explosion is usually a combination of flash, sometimes so intense that it causes involuntary eyeblink; a crushing energy wave that disrupts the system; a shattering noise that excludes all other ambient sounds; and a displacement whereby the victim is fortunate not to be hurled away to be injured by impact. The duration of an explosion may be a thousandth of a second; the event has passed before awareness is felt.

Circumstances surrounding the survival of individuals involved are usually more newsworthy than observations regarding the cause and progress of the incident. Nevertheless, they may serve to show the vagaries of the blast—another way of saying we cannot foresee the minor conditions that influence the local effects of a blast.

The medium through which a blast is transmitted is important. A person standing in the open air at a moderate distance from a small explosion may be almost unharmed (barring ear injury or wounds due to flying fragments). The elastic nature of the air dissipates the blast's energy as the square of the distance from the explosion. But in a nearly incompressible medium, such as water, the disruptive effect is so potent that the severity of an injury may be multiplied. It is probable that in some naval engagements a considerable number of the crew, after they were cast or had jumped into the water without great injury at the first explosion, were seriously hurt by subsidiary or following blasts transmitted through the water. Because the human body has about the same composition as the surrounding medium, it undergoes the same shock wave, and because this shock wave probably involves two-dimensional displacement (forward, up and down), it is particularly disruptive to the body.

B. Chemical Reactions

Let us return to the anatomy of an explosion. We offer two conditions that cover most instances:

1. Ignition of a combustible substance, (primarily the process of oxidation, i.e., combining with oxygen). Example: gunpowder.
2. Mechanical failure of a container so that an enormous volume of gas is released instantly. Example: a steam boiler.

The first is a chemical reaction giving off much heat, which causes enormous expansion of the gaseous substances created by the reaction; the second is wholly a physical expansion of an existing gaseous substance. Now gunpowder, dynamite, and so forth are substances that upon the initiation of a reaction are burned in the same sense that wood charcoal is burned in a fire (but infinitely faster). the combustion of lumps of charcoal takes oxygen from the air to create the gases carbon monoxide and carbon dioxide, thus:

C	and	O		CO	and	CO_2
carbon as charcoal		oxygen from air	react to form	carbon monoxide (a gas)		carbon dioxide (a gas)

But for combustion in an explosive, the oxygen is made part of the explosive itself. A much simplified example of a crude gunpowder would be a granulated mixture of charcoal, sulfur, and saltpeter (the chemical composition of the latter is KNO_3). When ignited, the powdered mixture burns with extreme rapidity, and we can visualize the reaction as being:

C	and S	and KNO_3	form	CO, CO_2, NO_2, K_2S, K_2SO_4, K_2CO_3
carbon	sulfur	saltpeter		(the first three are gases)

The proportions of the above substances are quite variable; the significant feature is that from a starting mixture of solids occupying very little space there is created a quantity of gas many times the volume of the solids. In the case of nitrocellulose, it may be assumed that a cubic foot of the explosive (about 110 pounds) would produce 10,000 cubic feet of gas. It is the very large volume of gas per unit weight of solid that makes the explosive suitable for a chosen effect.

The difference, then, between a fire (no matter how intense) and the chemical reaction of an explosive is that the combustible matter in the fire can burn only as fast as air (with its oxygen) can reach the flaming material, whereas an explosive carries its own instantly available oxygen.

Before we leave these explosion reactions, it is proper to consider another parallel between a fire and an explosion. We know that wood and paper will burn only if ignited, i.e., combustion is initiated by some outside condition. Perhaps a lighted match is applied; the flame is hot enough (about 1500°) to start the chemical reaction of the carbon of the paper combining with oxygen of the air. Once the paper is ignited, its burning can provide sufficient heat to ignite more paper. The fire becomes self-sustaining as long as air is given access.

Gunpowder, too, must be brought to a temperature at which the oxygen of the saltpeter can combine with the sulfur and charcoal. A spark, which is very hot, brought into contact with a grain of powder is sufficient to start burning; once started, the burning is self-sustaining. Man took the lesson of flint and steel for starting household fires and used it in the flintlock musket. The miniature explosion of the powder charge in the musket barrel was enough to cast a missile (here a bullet) far enough and fast enough to set in decline hand-to-hand combat. On a larger scale, it is artillery that, from medieval to modern times, has exploited the phenomenon of a man (the cannoneer) manipulating the explosive while he and the foeman may never see each other.

C. The Steam Explosion

We have used the above term as a convenient expression for the disruptive effects of the sudden release of gases that is not the result of the chemical reaction. Let us examine the attendant conditions of making steam. Liquid water, as is well known, evaporates from a saucer if left at room temperature. If the saucer of water is left in the sunshine and open air, evaporation is faster. The liquid is quiescent; we see no movement, and it may be many hours before the water disappears. A saucepan of water heated on a stove soon shows vigorous boiling and measurable loss of liquid. A thermometer placed in the liquid will register about 212°, and this figure remains constant while there is sufficient water to cover the thermometer bulb. We say that the boiling point of water at sea level barometer pressure is 212°.

The pressure cooker is a device in which steam is prevented from escaping except through a vent closed by an imposed weight. Steam, being unable to escape readily through the vent, builds up a pressure and finally lifts the weight, allowing some steam to be vented. The important feature of the pressure cooker is that as the weight impedes the escape of steam and causes a buildup of pressure, it deters the conversion of liquid water to steam. Yet the stove is steadily supplying heat to the vessel, and the contents have become hotter. This is the function of the pressure cooker: because the contents are hotter than they would be in an open vessel, food will cook faster.

The industrial steam boiler is an extension of the above principle. Water is confined in a strong-walled vessel and heated by a furnace. Venting is greatly restricted. The water must be very hot before it boils; the steam is at the same temperature as the water. The steam is eventually released in a controlled fashion to operate machinery or to supply heat to some process. New water is pumped into the boiler to replace liquid lost as steam. Figure 1 shows a rough relationship of boiler pressure and the temperature of the boiler's contents. Industry finds that for many purposes a high-temperature, high-pressure steam system is more efficient than a low-pressure system. Particularly in power plant generation of electricity there is an economic advantage to very high temperature and pressure.

A pound of water converted to steam in an open vessel (such as on a stove) will occupy about twenty cubic feet as a vapor. In a boiler such as one used in a public utility plant, the pound of water (as steam) would occupy less than a cubic foot because of the very high pressure. If the pressure were suddenly released, the steam would occupy, momentarily, a volume of more than thirty cubic feet. So it is that the accidental rupture of a boiler would result in the instantaneous creation of an enormous volume of very hot gas. There is no flame or spark, but great damage can take place from fragments of the boiler shell that become missiles. The expansion of the steam creates a shock wave transmitted to the surrounding area. Nearby temperatures approach that of the steam, but injuries from burns usually occur only within a short distance. The explosion does not initiate explosion of other boilers if these have not been injured.

The chief hazard of steam boilers is the failure of metal strength because of overheating which is usually due to not maintaining the flow of water to the boiler or from accumulation of foreign mineral matter dissolved and brought in with the feed water. (These substances may deposit in the boiler, and because they are poorer conductors of heat than the metal of the boiler itself, they lose less heat to the boiling water and the metal on which they deposit becomes overheated.) Overheating greatly reduces the metal's strength, and rupture of the boiler's shell

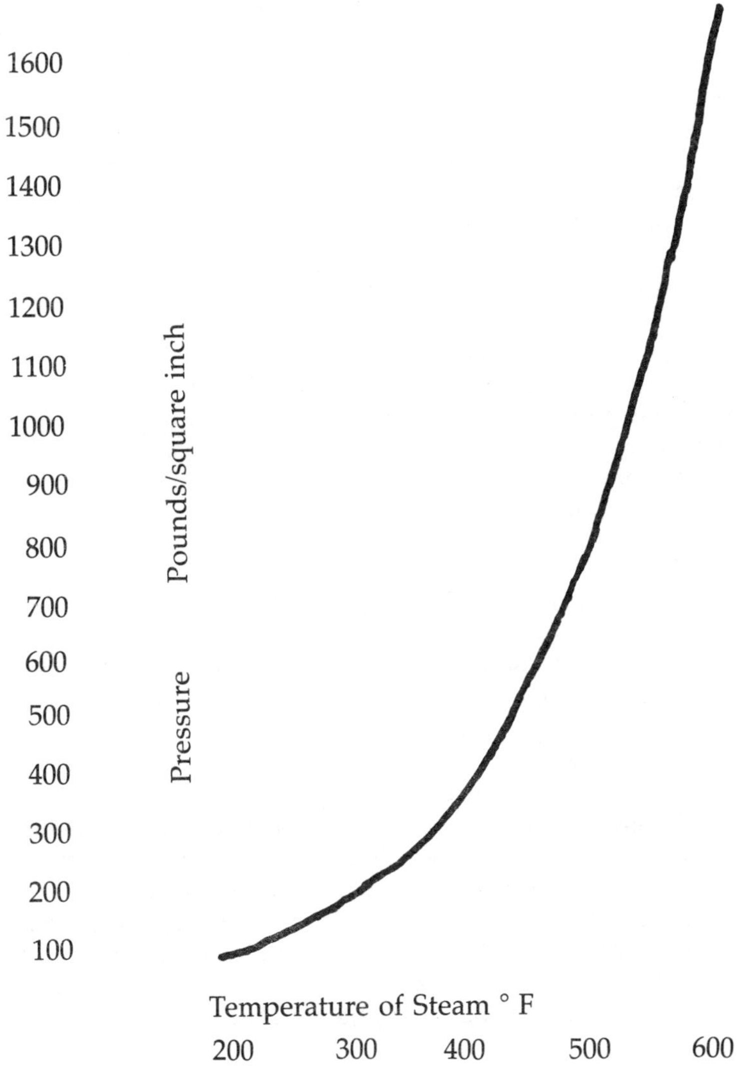

Figure 1. Boiling Point of Water as a Function of Boiler Pressure

can follow. The frequency of boiler explosions has dropped dramatically within a hundred years; regulatory bodies impose periodic inspections, and companies insuring the installations have exerted influence upon the boiler's use and operation, with far-reaching results. Technical societies have promoted improvement of design and construction.

Fortunately, few persons will be exposed to a boiler explosion, yet many may see its counterpart in a tree riven by lightning. Wood is a rather poor conductor of electricity, and lightning heats the moisture of the wood to the boiling point. There is no easy escape; the steam pressure builds up and bursts the fibers apart in spite of the considerable strength of wood. More often than not, a tree struck by lighting shows the effect of explosion rather than burning. Later we will refer to a steam explosion on a truly grand scale. But before leaving the phenomenon of steam explosions we reemphasize the principal features: (a) fire is not a necessary attendant, (c) no chemical reaction is present. A corollary of (c) would say that the explosion is not usually initiated by an outside effect or substance.

D. Indirect Damage and Casualties

Injurious effects following steam-boiler explosions sometimes equal or exceed those of the explosion itself. This has often been the case where the blast, although it carries no fire, produces a disrupting wave that may scatter burning combustibles from the furnace to other structures.

The question as to the extent of the indirect casualties arising from nuclear explosions over Hiroshima and Nagasaki must remain unsolved. Undoubtedly there were many radiation and blast victims. Not so evident was the subsidiary loss from conflagrations that followed. Some structures that withstood blast and afforded some radiation protection were consumed by fires fed by combustibles in the buildings, trapping many persons within the buildings.

A parallel circumstance attends some marine (and particularly naval) disasters in which survivors of an initial blast may have been blown overboard or forced to abandon the vessel without boats. In some instances the entire vessel seems to have been disintegrated and the duration of time between blast and submergence is academic.

Nor does the tragic end of the victims seem less bitter from ascription to one cause or another. The reader is reminded of the destruction of the dams in the Ruhr Valley (1944–45) by explosions in the impounded waters of the reservoirs. The explosions themselves caused little loss of life; the devastation by the released waters was enormous. We may assume that those planning and executing these operations had no primary objectives to inflict casualties, but rather sought the disruption of the economy depending on the dams. We say this because in the case of the usual air raid (by either combatant) the prospect of civilian casualties is accepted as an incremental gain along with the destruction of any military or industrial targets.

E. Difficulty in Measuring Effects of an Explosion and of Determining the Cause

If it were possible to tabulate explosions (other than natural events) of the past two hundred years in a sequence of increasing amount of explosive involved and to juxtapose the resulting damage (injuries or material destruction), even the casual reader would note a very imperfect correlation of explosion size and damage inflicted. So much depends on the circumstances: the type of explosive, the environment. For example, the powder mill and the fireworks factory have somber histories; employees are said to be in very hazardous occupations.

Yet as technology sought for and derived (imperfectly) a correlation, several criteria seemed valid: (a) displacement of solid matter adjacent to the site, e.g., by cratering; (b) the damage or breakage apparent in concentric circles about the site; (c) shock waves felt or measured seismically; and (d) audible sounds reported by distant persons. It may be summarily stated that none of these criteria has been quite satisfactory; even when two or more of the factors are involved, the correlations present important deviations.

Take the matter of reported noise of explosions heard at distant points. Close to the immediate site, the intensity is high and many persons will unconsciously note the incident and consciously observe the time. As we move away from the site, fewer persons will think the noise is related to a significant event and even fewer will note the time. Finally there will be only a few sensitive and observant individuals upon whom is registered an unusual sound. If one were to plot the observations of recorded sound against distance from the site, we might have a grouping as shown in the diagram below, where the concentric circles represent distance. But it seldom works out this way; in likelihood the plot will be an amorphous shape (figure 2) or, exceptionally, like figure 3. Persons within the shaded area failed to note unusual noise. Obviously such distortion of the pattern requires study. Often the explanation lies in differences of elevation and the interposition of hills or woods. Also important are atmospheric conditions, particularly temperature and humidity. Sounds in air can be absorbed, reflected, refracted. Histories of disasters have sometimes told how a distant locality received tokens within minutes while nearer-by towns were unaware of the event and learned of it much later by word of mouth.

It may come as a surprise to many people that they have seen a shock wave. But the phenomenon is frequently registered in motion pictures. Observed from above or at a distance, the detonation of an explosive will show the extremely fast passage of a wave outwardly from the explosion site. There appears to be a distortion of the ground features against which the picture is taken; these features seem to waver.

The air through which the features are seen is compressed and rarified. The effect is akin to the distortion that can be seen on a hot summer day during which the sun's heating of the air causes objects at a distance to appear to be dancing.

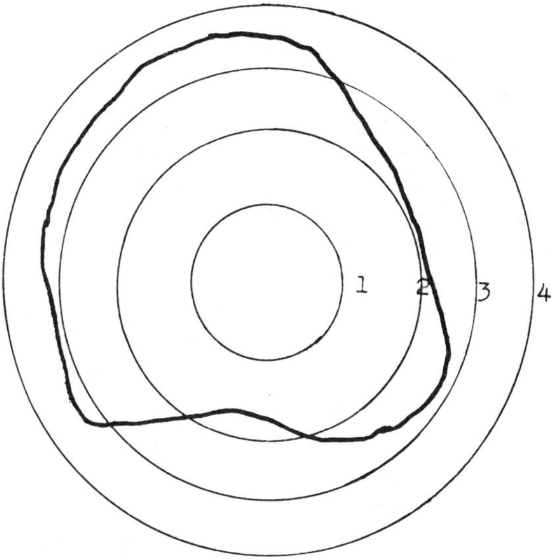

Figure 2. Showing usual pattern of explosion sound heard at a distance

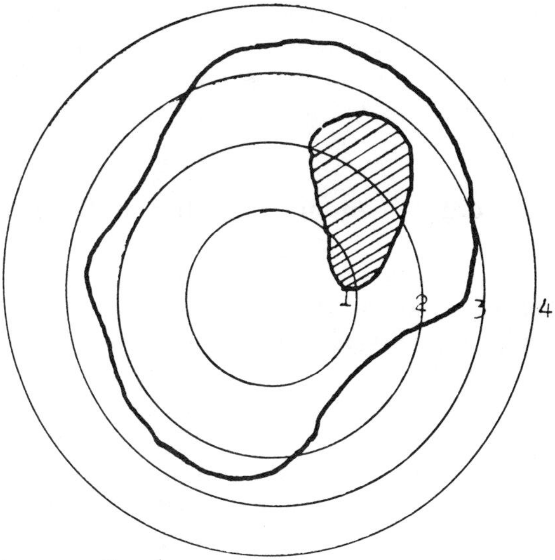

Figure 3. Showing exceptional pattern where an isolated area did not hear the explosion.

Shock waves passing through the ground are likewise cousin to earthquake motion, although they are of far less magnitude; instruments can detect and measure minute shocks of explosives. Unlike earthquake tremors, man-made and accidental explosion shocks are shallow. A powder-mill blast, although huge in terms of explosive involved, takes place at or near ground level; much of the energy is dissipated in the air. A lesser amount of explosive detonated in a deep hole may produce a distinct seismic effect. Early gunpowder-makers built structures with light wooden roofs and perhaps one side that would be flimsy framework. Eventual explosions were expected, but the substantial masonry was relatively uninjured.

There has been much evidence of the relative forces of air blast and of ground tremor obtained from wartime bombing of industrial plants that resulted in very heavy damage to surface structures except stacks such as power plant chimneys. These tall structures would seem to be very vulnerable to air blast and ground shock waves. However, seldom were they destroyed, except by direct hit. We may conclude that the lateral displacement at the base was small and that the air blast was not potentially destructive to these objects.

F. Confinement and Duration of the Explosive Reaction

Destructive effect of explosions is enhanced by confinement of the explosive. Compare the explosion of a firecracker per se and the ignition of the contents in the open air, producing merely a brilliant flash.

We spoke of an explosion as an instantaneous event; in actuality it is not. The time lapse is measurable, although it is only a small fraction of a second.

Suppose, for example, that a thousand tons of TNT are closely packed in a cube measuring thirty feet on a side. If this mass is detonated by a fuse at the center of the cube and there is no displacement of the mass, as much as a thousandth of a second would elapse before all the TNT underwent chemical reaction. This seemingly trifling amount of time is very important. The more nearly instantaneous an explosion is, the more drastic its effect. There is less chance for dissipation of energy by radiation and absorption in the yielding surrounding matter. In some applications, such as mining, this very excess of explosiveness can be said to defeat itself. The ore in the immediate explosion area may be pulverized, melted, vaporized. All such conditions result from the absorption of energy that was primarily intended to break ore into manageable pieces and to displace the broken material into position for handling. This has led to the concept of detonating charges in sequence so that the effect is that of a "long" explosion. The individual explosions are separated by hundredths of a second. The effects are pronounced. The initial shock provides fracture of the gross mass; the succeeding charges move broken material. The latter—when the procedure is properly done—is an act of pushing rather than throwing. Of course in accidental explosions no such condition prevails. When there is little confinement, the explosive force is directed upward and outward, and little cratering results. The damage is localized, except for the effects of air shock, such as broken windows at a distance. Where there is confinement, the blast effects become more and more selective. Weakest parts of the structure are escape avenues; a larger part of the energy is directed horizontally and cratering may be more evident. All this may be difficult to discern where entire structures that were the seat of the explosion disappear.

The unexplained vagaries of the conditions in explosions may be exemplified by an incident:

An interurban trolley car was subjected to a very severe explosion, and heavy casualties among passengers resulted. A wagon had been engaged in the delivery of (amongst other things) a box of dynamite. The box fell unnoticed from the wagon and landed across the trolley tracks. Shortly afterward, the trolley car ran upon the box; the box was crushed, and an explosion followed. Subsequent inquiry involved the explosives manufacturer, the express agency, and the traction company. Here is a rather typical circumstance that illustrates the difficulty in determining cause when there is imperfect evidence and placing responsibility.

The explosives maker maintained that particular package involved was no different in quality from other dynamite of the same lot and that instability from imperfect manufacture was not the cause of the tragedy. The express company claimed that the transport of explosives within the city was sanctioned by custom; the unnoticed falling of the package from the wagon was attributed to carelessness on the part of the driver; it was dusk and near the end of his tour of duty. The tailgate may have been improperly fastened, or perhaps there was an extraordinary jolting as the wagon passed over the tracks. The trolley motorman (who was killed) would seem culpable in some degree for not noting the box in his path. We have said it was dusk at the time of the accident; the damaged condition of the trolley after the event did not permit conclusions as to whether the trolley's headlight had been functioning. The traction company had to assert all equipment was in good operating order or open itself to a charge of negligence. We can only conjecture why the motorman ran the car upon the box: (a) he did not see it because of a nonoperating headlight or his poor eyesight or (b) he saw the object in the trolley's path but assumed it was unsubstantial and would not impede the car and would be brushed aside by the fender.

What lessons can be derived? We see a number that, with hindsight, are applicable:

1. The initiation of the explosion could have been brought about by any of the following: (a) part of the contents of the box were unstable; (b) crushing of the box and its contents resulted in traces of free nitroglycerine; (c) after the package and its contents were crushed on the track, the passage of the current from the trolley's motor to wheel and track resulted in a spark; or (d) as the box was crushed on the track, a nail in the debris dragged against the roadway, producing a spark.

2. The transport of explosives, particularly in built-up areas, should have been subject to much closer regulation.

3. There was demonstrated the need for very high-stability explosives that would not be detonated even by crushing (there was no statement that blasting caps accompanied the dynamite).

4. The utility would seem to share the fault. Equipment inspection and employee training are needed.

This example is offered as typical of many explosions where, though evidence is lost in the accident, conclusions may be drawn that, applied, would lessen frequency of such catastrophes. It was noteworthy that the investigators obtained no crucial evidence from eyewitnesses, persons nearby the scene. The case presents most of the classic conditions of catastrophe:

1. No warning or premonition existed.

2. Persons who might have given important information are lost.

3. Physical evidence is destroyed or distorted by the blast itself.

4. Trivial or bizarre circumstances have contributed.

In retrospect, it can be stated that the incident agrees with many others in that the tragedy was avertible.*

* See also incident at Cali, Colombia, site of a great catastrophe.

2. Explosions Sensitizing International Conflict

An explosion that in ordinary circumstances would be a regrettable catastrophe may, in the framework of singular times, threaten to become an international holocaust. Public tempers are inflamed; the causes of explosions, so often impenetrable by quiet investigation, are imputed to the deliberate ill will of an opponent. Within modern historical view there have been at least two such incidents, one of which was an important motive for declaring war.

The inflammatory conditions in Algeria in the 1950s prevented reasonable discussion among all parties, whether of European, colonialist, or nationalistic inclination. When, then, the Egyptian munitions ship *Star of Alexandria* blew up in the Algerian port of Bône, the ascription to sabotage was prompt and violent. The loss of almost a hundred of the crew and dock workers could be a small toll compared to that brought by to open warfare and clandestine reprisals.

Sober consideration of the *Star of Alexandria* incident confirms our belief that the court of inquiry had in its hands the means of averting immense disaster. The difficulty of assessing the cause of any explosion—particularly in circumstances where persons immediately present cannot give evidence—is foreseeable. Sabotage was a distinct possibility; the destruction was rejoiced in by some parties, and if these parties did not cause the explosion, they at least wished they could have done so. To this writer, the accidental nature of the catastrophe cannot be dismissed. An atmosphere of violence, a hardly disciplined work force, and the inexperience of crew and authorities in handling large quantities of explosives point to the likelihood of accident. The court's finding was that the explosion was accidental but avertible.

The year 1898 saw an incident that must be said to rank with that in Algeria but that led to events both more prolonged and wide in scope. The USS *Maine* was in the harbor of Havana, Cuba (then under Spanish jurisdiction) and was destroyed February 15 by an explosion. Over 250 of the ship's crew were lost. An investigating group found that an exterior mine had detonated the ship's magazines. Spanish authorities conceded that the harbor was mined but denied any complicity in the incident.

Public sentiment in the United States was for a war with Spain, and declaration of hostilities followed in two months. Without ascribing the cause of war to the *Maine* incident, we may say it triggered open conflict.

About twenty years later, the wreck of the *Maine* was raised from Havana harbor and was sunk in deep water. No detailed results of the examinination of the wreck were made public, but it was inferred that the original verdict of "destruction by an exterior explosion" was valid.

With the lapse of years and the subsidence of inflamed public opinion, we may make two guesses at actual circumstances:

(a) A mine, part of the Spanish harbor defenses, was detonated accidentally or without authorization.

(b) An explosive device was planted and detonated under the hull of the *Maine* by an insurgent group favoring open hostilities.

It is highly improbable that a deliberate, hostile act was performed by Spanish authorities. That nation was quite unprepared for a war with a power like the United States, and even if Spain had wished to make war and had been better prepared, the approach is without reason. The results of the unequal conflict marked the precipitous decline of Spain as a colonial power. The casualties in this war were very heavy for the Spaniards, light for the Americans, although the latter suffered greatly from campaign illnesses.

The total incident may be viewed as an unexpected disaster whose reporting led to or incited yet more serious effects. We may guess that had the explosion not occurred, the question of Spanish rule (or independence therefrom) would have been resolved with far fewer deaths for both major nations.

3. Air-Separation Plants

A Danger Defused

Air-separation plants,—which are capable of taking ordinary air and separating it into its major components, nitrogen and oxygen—are part of modern technology. They can make this separation with remarkable effectiveness, and the process has served industry well. High-quality oxygen is used by the ton in steel making, and lesser amounts are used in welding and the cutting of metal. What of the nitrogen gas, produced in amount four times greater than that of oxygen? (Air is only a physical mixture of nitrogen and oxygen.)

Within this century processes have been developed that cause nitrogen to unite with hydrogen to form ammonia. The reaction is represented in chemical symbols: $N_2 + 3H_2 = 2NH_3$, or 1 volume of nitrogen gas and 3 volumes of hydrogen form 2 volumes of ammomia. Ammonia is used in enormous tonnages for the preparation of fertilizers. Furthermore, it is possible to make nitrogen-oxygen compounds, amongst which is nitric acid, used in explosives manufacture as a primary ingredient in nitroglycerine and TNT (trinitrotoluene).

Since oxygen is the element in conventional explosives enabling them to burn exceedingly fast, one might expect that the production of oxygen as a concentrated gas would be an extraordinary hazard. Air-separation plants present an anomaly in that the process and one of the products, though seemingly extrahazardous, have proven relatively predictable in safe performance.

The term "relatively" creeps in and was appreciated only after the occurrence of accidents that long defied explanation. In these untoward events, the conditions present at an incipient explosion were altered by the explosion. Gradually it had been established that foreign combustible matter was present, but the means of its introduction were not evident. Finally it was realized that minute quantities of material, chiefly hydrocarbon, were present in the air intake and became concentrated in the process equipment. This concentration awaited a condition favorable to reaction and thus was an ever-present danger. Once this situation was appreciated, prompt and simple remedies were applied: measure the accumulation of foreign reactive material and withdraw this at suitable intervals.

The example of air-separation plants illustrates the shattering of complacency by unforseen minor episodes, recognition of a major hazard, and the use of effective countermeasures.

4. The Biggest Bang on Earth (So Far)

An Analogy to Steam-Boiler Explosions

Noise is a disturbance of the air, and we rightly associate the loudness of an explosion with the size of the explosion (quantity of explosives). The louder the noise, the farther away it may be heard. Blasts of conventional explosives have been reported at a distance of one hundred miles. But it is a tricky matter, and in some cases there are zones of silence at lesser distances. The most gigantic terrestrial explosion recorded is that of the island of Krakatoa in the Sunda Straits (between Sumatra and Java).

The vicinity of Krakatoa has a long history of volcanic activity; the principal island itself was probably born of an eruption. In August 1883, unusual activity of its volcano was noted; many sightseers were attracted. On August 27, the eruption climaxed with explosions that transformed an island whose elevation was 300–1400 feet above sea level into a cavity extending a thousand feet below sea level. There is no accepted estimate of the solid matter displaced, but the unleashed forces would dwarf those of man-made hydrogen bombs tested in remote Pacific areas. The sky was darkened at Batavia (now Jakarta), one hundred miles away; the sound was heard over three thousand miles away. One of the peculiarities of the sound waves resulting from the explosion of Krakatoa was that the noise of continuing blasts virtually ceased to be heard in some nearby areas. This was purely a local effect, and the phenomenon was attributed to the clouds of dust that absorbed the sound waves. Altogether it was reckoned that the noise of the initial explosion was heard (or could have been heard) over one-thirtieth of the globe's area.

An analogy to a steam boiler seems reasonable. If sea water penetrates to a zone of molten lava that is under a gentle but irresistible pressure to escape from beneath the earth, the steam immediately produced might push some lava upward. But as this lava meets other water and is cooled, it will solidify and stop all flow as would a plug. Pressure builds up behind the plug as the deep-seated forces try to move the lava along. The steam pressure mounts, and finally the steam and lava are ejected in a tremendous blast. These conditions are not part of the ordinary volcanic eruption, in which little water is available for contact with bottled-up fluid lava. Until man detonates several hydrogen bombs simultaneously, we are not likely to know anything like Krakatoa or Coseguina, and in the event we do, we will have nothing to record as personally witnessed.

If we were to seek a classic example of an explosion reaction in a volcano, perhaps we could do no better than to cite Krakatoa's spectacle. So enormous was the initial impulse wave that its effects were detected on the tide-gauges of Britain and France. From observations there and at many intermediate points, it was calculated that the wave traversed at speeds up to four hundred miles an hour, depending on the depth of water en route (the waves traveled at higher velocities in greater depths). It was reported that the greatest wave was about fifty feet high. It was also estimated that the initial wave struck nearby islands at speeds exceeding a hundred miles an hour, clearly making the calamity one of destruction as well as inundation. Thirty-six thousand persons perished, most fatalities probably due to sea waves inundating shores within a radius of a hundred miles. Dust was hurled into the air to an estimated height of over twenty miles and produced lurid sunsets for a year.

Clearly, a distinction must be made between eruption and explosion. The great killers, Etna and Vesuvius, discharged material over long periods of time. Etna (in Sicily) was reported to have caused 100,000 casualties before its eruption in 1669. Mount Pelée (Martinique) in 1902 claimed 40,000 lives. Probably Pelée's *nuée ardente* caused most deaths by asphyxiation and burns. The disruptive effects of a typical explosion were not noted. We must be wary of such figures because the deaths are not necessarily immediate attendants of an explosion. It is true that minor explosions may have accompanied the pouring forth of overwhelming lava flows and suffocating gases, but the eruptive phenomenon is not the classic picture of an explosion, a virtually instantaneous release of energy.

Katmai (Alaska) and Bezyiamanny (Kamchatka) exhibited enormous releases of energy, but it is inferred that the activity took place during measurable time. Bezyiamanny was active from October 22, 1955, to March 29, and then erupted in a gigantic explosion on March 30. Curiously, the sound was not heard at great distances, as was the case of Krakatoa, but an air wave circled the world one and a half times. The region is quite desolate; no casualties resulted, although the physical features of the site were vastly altered. The amount of change due to the explosion itself was not determinable.

Tambora (East Indies) in 1815 underwent what may have been, in terms of displaced matter, a greater eruption than Krakatoa. Ten thousand persons were killed outright; many more died indirectly, as by starvation. However, the volcano's activity continued over several days and it is impossible to state the effects due to the paroxysm whose sound was heard at distances of upwards of nine hundred miles.

Coseguina (Nicaragua) started erupting on January 20, 1835, and threw out immense amounts of ash; a climactic explosion took place on January 22, and was heard for hundreds of miles. Intrusion of sea water to reach molten lava initiated the explosion, which ejected enormous amounts of lava subdivided into solid sand-like particles by the steam. Eruption ceased on January 23. The energy involved, as measured by displaced matter, places Coseguina amongst the foremost of violent eruptions. But the uncertainty as to its distribution in time (over a period of four days) subordinates the event to Krakatoa's. The area suffered blanketing with sandlike deposits, but unlike Krakatoa, which demolished an island, Coseguina was satisfied to make a few changes to the nearby landscape. Perhaps Coseguina may make modest claim to being the second most intense noise event.

5. Energy in Some Explosive Volcanic Eruptions

Here we enter a field of open dispute. Some scientists have estimated the forces involved by computing the amount of solid matter displaced during the eruption, at times arriving at figures of *several cubic miles.* To move such a mass of rock requires energy beyond our usual concepts. A cubic yard of rock might weigh about two tons; a cubic mile of such rock would have a weight of about ten billion tons. The combined railroads of the United States hauled freight in the amount of one and a half billion tons in 1970; the trucking industry estimates that trucks conveyed nearly two billion tons in that year. We are dealing with enormous amounts of volcanic debris, but there is dispute regarding the quantities. There is the distinct possibility that voids attributed to the ejection and dispersal of material as the result of an explosion may in part be caused by the subsidence of the area in the immediate vicinity of the volcano. Those persons who estimate the quantity of debris by observations of ash clouds are severely limited by being unable to measure the composition of these clouds near to their source in distance and time. Admittedly huge quanties of pumice or ash were noted on the seas surrounding the vicinity of Krakatoa, as were deposits on the seabed itself.

With all these shortcomings, at least one estimate places the energy involved in Krakatoa's paroxysm as 1×10^{25} ergs. For comparison we may note that a megaton of TNT is estimated to have energy of 4.2×10^{23} ergs. The energy released in Krakatoa's destruction may then be said to have been equivalent to that of 24 million tons of TNT (24 megatons).

The distance that noise of an explosion travels is probably a good indication of the intensity of the blast. If this is accepted, we may rank Krakatoa, Tambora, and Coseguina as immense but uncritically measured explosions.

It should be noted that the energy in the outpouring of molten lava may greatly exceed that ascribed to explosive eruption and is distributed over long periods. The energy ascribed to several explosive eruptions has been estimated:

Tambora	8.4×10^{26} ergs	equal	to	2300	megatons	TNT
Coseguina	4.8×10^{26} ergs	"	"	1140	"	"
Bezyiamanny	2.2×10^{25} ergs	"	"	52	"	"
Krakatoa	1.0×10^{25} ergs	"	"	24	"	"

We may question the significance of these figures but cannot doubt the lavish expenditure of energy by nature.

6. Beneficent Explosions

We are so accustomed to world news of disasters that we may merely shudder at the thought of "another explosion." We do not expect to modify such morbid reactions, but we can find interest and perhaps delight in some explosions with very different impacts upon our surroundings. Of course we could include the minuscule, such as expanding cereal grains to make breakfast food, or the exciting bang of firecrackers or more sophisticated fireworks. But there have been cataclysmic explosions performed by man causing no casualties and in some instances certainly saving lives that otherwise might have been lost.

Not least impressive in such incidents is the complete reversal of attendant conditions in the vast number of recorded disasters incident to explosions. Among these beneficent, planned explosions we see the following:

- Precise foreknowledge of the event
- Knowledge of the cause of the explosion
- Knowledge of how much explosive was involved
- The ability, in part, to direct explosion's effect
- The opportunity to record the physical conditions of the explosion and its effect upon surroundings.

Others may occur to the reader, but the above knowledge removes the unresolved aspects of the conventional disaster.

The use of explosives in mining operations to recover ore may be said in a broad sense to be beneficent, albeit the immediate motive is profit. The disposal of explosives to remove possibility of use by a hostile force can have some claim to merit in man's affairs. There are examples of both of these where the quantity of explosive involved was stupendous.

We will now discuss an event where the primary intent was to better the condition of lives, possibly to save lives. No monetary profit was used to justify the expense; the explosive itself was common to civilian life; the operation was conducted by civilians. It caused no casualties. It embodied all those features of beneficent explosions: time, place, and amount were prestated and instrument recordings of effects enabled comparison with predicted effects.

A. Ripple Rock Detonation as an Example of Beneficent Explosions

A navigable passage exists in the waters between the mainland and outlying islands from Puget Sound to the Gulf of Alaska. The protection from direct Pacific Ocean forces is almost continuous throughout its length of eight hundred fifty miles. Ripple Rock was located in this passage about a hundred miles from Vancouver. It was submerged at a depth of only a few feet at low tide; it created great tidal disturbances. Actually, the Rock represented two summits in a hogback 3000 feet long and 350 feet wide. The long history of the Rock as a hazard to navigation is studded with proposals to neutralize this danger. Curiously, strong opposition to elimination of the Rock came from proponents of a railway bridge across the narrow strait; it was proposed to use the Rock as footing for bridge piers. Construction of the bridge never materialized, and, in the light of later experience, there is serious doubt as to whether the bridge could have been completed without enormous and unforeseen costs. The chief difficulty in building of piers could be ascribed to the turbulent tidal flows, which would reduce effective working time to two short periods each day.

The solution finally accepted for removal of the obstruction was to attack the Rock from below, rather than above. This entailed a tunnel starting from a point on the east side of the channel (the mainland) and branching beneath the Rock, and honeycombing the mass with chambers to receive explosive. (Figure 4 is a cross section of the passage.) The intent was to disintegrate the two peaks of the hogback and to disperse the debris over a wide area of the channel. A total of almost fourteen hundred tons of explosive was placed and was detonated as a single blast.

The operation was a technical success. The channel obstruction was reduced to give a minimum depth of water of forty-five feet. The fierce, erratic tidal disturbances were notably lessened. The project yielded subsidiary benefits; instrumentation of the area gave valuable information on air shock and seismic effects.

In one respect only can the experiment be viewed as short of satisfactory: Was the amount of explosive used optimum? The depth of water and the strong water movement prevented adequate examination of the new channel floor at the Rock. The perfectionist would like to know whether a lesser amount of explosive, arranged in another manner, would have yielded equally good reduction in height of the obstruction and an excellent dispersion of the debris. We shall not cavil at this magnificently beneficent explosion, whose only casualties were channel fish.

B. Blasting to Provide Fill for a Railroad Causeway

The route of a major railway crosses the shallow Great Salt Lake in Utah. A plan was proposed whereby several miles of wooden trestle would be replaced with a causeway to carry the tracks. Building the causeway necessitated an enormous amount of rock and dirt. To provide this fill, engineers blasted material from a cliff near the lake. A series of great blasts was made, culminating in January 1958, in an operation using about 1,060 tons of nitrate fertilizer and dynamite, which made over three million cubic yards of fill available. Controlled explosions like

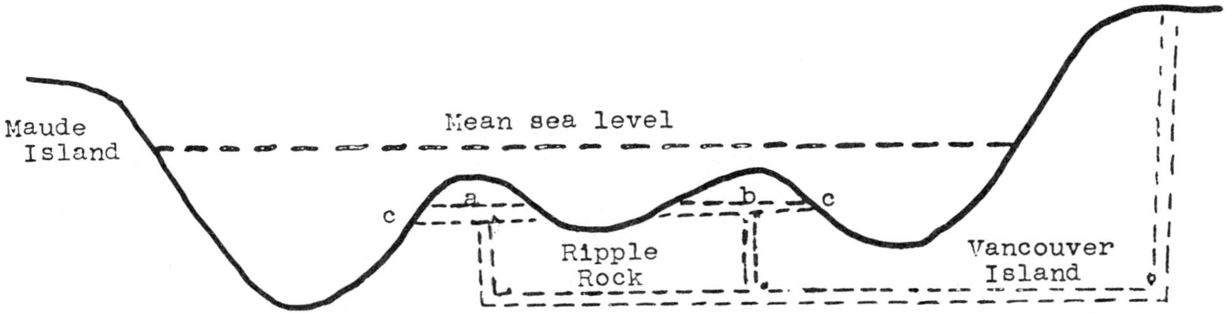

Figure 4. Cross section of channel between Maude Island and Vancouver Island. Sites a, b honeycombed to receive explosive. New channel minimum depth indicated by c-c.

this without resulting tragedy yield valuable engineering data while performing tasks that would otherwise involve great expense and considerable labor.

C. Peacetime Disposal of an Enemy's Facilities and Munitions

What to do with the German naval base at Helgoland was a question confronting Britain at the end of hostilities in 1945. Helgoland was strongly fortified during World War II and was adapted to serve U-boats by providing pens considered secure from aerial attack. The huge store of munitions (conventional naval and artillery ammunition, depth charges, torpedoes, and aerial bombs) posed a threat to civilian occupation. The enemy facilities were to be neutralized or destroyed; the British planned to solve both problems by using the munitions internally, i.e., to detonate the explosive stores to demolish the facilities. On April 18, 1947, the Royal Navy conducted an explosion of over three thousand tons of high explosive, consuming the stored munitions and completely neutralizing the martial facilities. Hopefully, Helgoland will revert to its former attraction as a beautiful vacation spot and bird refuge.

7. Coal-Mine Explosions

An Ever-Present Tragedy

Since the advent of the Industrial Revolution, which required enormous quantities of fuel, the mining of coal has accepted the attendant human losses almost with fatalism. The destruction of property has been large, the loss of lives disgraceful.

Aside from the hazard of a roof-fall in the mine, which might injure the worker, the miner recognized the release of inflammable gas from the coal seam itself. Open lights (candles and oil lamps) were the usual illumination; the miner became aware that a dangerous gas concentration had built up. The flame might waver or flare above the wick and perhaps have a less yellowish aspect. The miner was fortunate in that the gas burned in the flame, thus being consumed harmlessly, although it depleted some of the oxygen supply. It represented the combustion of a local, rich gas-air mixture, not a widely dispersed, explosive mixture.

Methane, which is the hydrocarbon CH_4, is the principal component of coal-mine gas and natural gas. It is odorless, colorless, tasteless, and about half as heavy as air. It can burn cleanly to give carbon dioxide and water vapor. Weight for weight, it has one and a half times as much energy as gasoline and three to four times as much as TNT. It is easy to see that this dread substance held a shadow over the lives of coal miners. We said that methane burns in air; this requires further explanation. It burns readily only when its concentration in a mixture with air is more than 5 percent and less than 13.5 percent. These figures are known as the explosive limits of methane.

Two countermeasures were known to offset the hazard of methane explosions: detection of the gas and removal of the gas before large explosive concentrations were built up.

The safety lamp, child of Sir Humphrey Davy's fertile brain, went far to solve the first problem. Davy's device was a lamp wherein the flame (from candle or oil) was shielded by metal screening. Air could enter through the metal gauze and support combustion at the wick; the hot gases of combustion could escape through vents. Any methane present in the air entering through the screen would cause the same changes in the flame noted above. The methane burned harmlessly inside the lamp; the flame never passed out through the screen's meshes because the metal gauze absorbed heat from the flame and radiated this heat at a much lower temperature than the flame. Loss of some light due to the screen was a low price to pay for the safety provided by this lamp.

The second feature was to provide better ventilation of the mine, thereby achieving dilution of dangerous concentrations of methane and continuous removal of contaminated air. Mine ventilation today is accepted practice both for safety and bodily comfort of the miners.

Despite the above measures, the number of fatalities in coal-mining operations remained high throughout the world. A subtly hazardous condition was present. Much coal is friable, i.e., easily crushed; mining often produces much coal dust. The dust is highly inflammable and, under certain conditions such as when suspended in air, is tremendously explosive. Engineers seeking causes of disastrous explosions in certain mines were convinced that minor blasts had caused huge amounts of dust to be shaken into the air of mine passages, thereby creating highly inflammable mixtures representing a danger that outweighed that of the coal gas. Three routes to lessen the hazard are seen: good housekeeping, i.e., do not allow dust to accumulate; wet the dust to make it less likely to form a cloud; disperse a nonflammable dust (rock powder), which in effect dilutes the explosive material to a safe limit. Any of these measures may be taken today in coal-mine operation.

Dust explosions, whether the material is coal, dried milk, or grain, present the same phenomena. As combustible solid particles are reduced in size, they present ever greater amounts of surface for oxygen of the air to react upon. Contact of flame with a particle immediately heats the particle to its ignition point. Ignition releases more heat; the process becomes self-sustaining and explosive.

In the first quarter of this century there were several disastrous explosions, each involving at least two hundred fatalities. We cite three, together with observations made by investigators:

1. May 1900, Winter Quarters, Utah; 200 fatalities. There were said to be gassy conditions in this mine. It was noted, however, that there was much dust and that excessive blasting powder may have been brought into the mine.

2. December 1907, Monogah number 6 and number 8, Mono gah, West Virginia; 366 fatalities. Explosion may have been initiated by loaded coal cars breaking away and piling up, thereby creating a cloud of dust that could have been ignited by the arcing of a broken electrical cable.

3. December 1907, Darr Mine, Jacobs Creek, Pennsylvania; 240 fatalities. This was not considered a dangerous mine; open lights were permitted.

Three catastrophes, and a lingering doubt in each case that the tragedy was inadvertent. Among the hundreds of mine explosions in this century, there is a persistent thread of relaxation from a rigid pattern of discipline, on the part of both management and miner. Increasing public awareness has led to federal and state safety regulations and inspection, and this had undoubtedly enhanced mine safety. Latter years have shown many explosions and fatalities, but the ratio of tonnage mined to miner casualty has risen steadily in underground operations. Safety measures and safer equipment, impartial inspection of operations and enforcement of regulations, and the greater use of mechanical methods of extracting coal seem prime factors. Mechanism reduces the number of miners per gallery, i.e., the workers are distributed over a wider area and an explosion in one part of the mine may involve only a small portion of the total working force.

Despite all legislation and good will, coal mining yet presents hazards, and recurring incidents are a clear token that vigilance can never be relaxed. American mine disasters can find many counterparts throughout the world. Coal-mine explosions are good examples of phenomena where conditions are well understood but the arrival at total safety is peculiarly susceptible to human error.

8. Dusts: A Hazard Belatedly Recognized

Careful analyses of coal-mine explosions show repeatedly that coal dust can be a contributing factor, and often the dominant element, in the destructive effect, for the result of any gaseous explosion is to produce a cloud of inflammable particles from coal dust that lies inertly on ground and equipment.

Dealing with the cloud of coal dust can have its counterpart in the handling of any finely divided carbonaceous matter, whether it is flour or milk, or formed by attrition, as may be the case with transfer of dry grain. The size of these explosive events is small compared to coal-mine disasters, but their collective toll is distressing, and if discussion of specific instances here is not warranted, yet they serve to prick our conscience. Technology and society recognize that some explosions are preventable; the latter party tends to make the accusation that the recognition is belated. For example, the explosion in a grain elevator connotes two conditions: an inflammable dust and a means of ignition. To allay the one and to obviate the other may be both difficult and expensive.

9. Munitions-Ship Explosions

A. Halifax, Nova Scotia, Munitions-Ship Explosion

On December 6, 1917, occurred what can be regarded as the prototype of munitions-ship disasters in the vicinity of civilian activities. We say prototype because the repetition of such catastrophes at far distant points seems to reflect harshly on our ability and willingness to apply hard-won experience.

The incident in Halifax waters may be told briefly. The French munitions ship *Mont Blanc,* carrying twenty-five hundred tons of explosives was moving on a course approximately parallel but opposite to that of the Norwegian vessel *Imo,* which was loaded with grain. Mistaken or ignored signals resulted in collision, which ignited inflammable liquid in drums on the *Mont Blanc*. Failing to extinguish the fire, all personnel abandoned this vessel, which continued to move toward a congested land area. The ship exploded, causing two thousand fatalities and injuries to nine thousand other persons. There was immense property damage. The time of year (December), the destruction of communications, and the disruption of services placed such a great burden on rescue and relief that suffering was intensified.

The immediate cause of the disaster was the collision of the vessels; why they collided remains a mystery. The master and harbor-pilot of *Imo* were lost; responsible personnel of the *Mont Blanc* (including a harbor-pilot) survived to give what is—understandably—rationalized evidence of events and action.

Observations and Comment on the Situation

The presence of a harbor-pilot on each vessel accents the mystery of the courses of the ships. We can ascribe the action or nonaction of the *Mont Blanc's* crew to panic or ignorance. The hasty escape after an unsuccessful fight to control the fire was a recognition of imminent danger of explosion. The failure to direct the vessel toward a sparsely occupied portion of the shore reflects panic. It may never have occurred to the master to sink the vessel, nor can we be certain that he was aware of means and had had instruction for such emergency. A naval vessel had witnessed the accident and had dispatched a party to board the *Mont Blanc* and fight the fire. All the boarding party were lost, so it is impossible to say whether flames at the time of abandonment by the *Mont Blanc's* crew had definitely prevented change of the ship's course or blocked access to flooding controls. There was an interval of twenty-one minutes between the collision and the explosion, during which time many persons had gathered near the waterfront. It is impossible to say how many of the casualties were among those persons solely in the role of spectators.

Retrospect

Leaving out the unexplained course of the *Imo* (as in many explosions, firsthand evidence was destroyed), in considering this incident we look for means of averting or lessening the horrors that are potentially present where huge amounts of explosive are at hand. As these means are enumerated, they do not seem so far beyond practical accomplishment that they may not be readily adopted as obligatory measures.

(a) The large-scale handling of munitions should be done at locations well removed from civilian centers.

(b) Vessels carrying munitions should be marked prominently and all shipping in the vicinity should be made aware that a munitions carrier is present.

(c) Crew personnel should have intensive fire-fighting instruction and should be able to operate flooding valves. The seacocks of munitions carriers should be designed to cause prompt submergence of the hull.

(d) The presence of persons as spectators in a potential disaster area should be controlled.

We shall see how in a later war, some of these lessons were ignored with disastrous consequences.

B. Explosion of the Ship *Fort Stikine*

Bombay, April 1944

This event is in a particularly sad category, as it involved many victims without their knowingly playing a part. The steamship *Fort Stikine* was a munitions carrier, yes, and the crew were aware of the inherent dangers, just as the crew of a warship accepts attendant hazard as a condition of the service. Many of the citizens killed or injured in this explosion possessed awareness without appreciation of latent danger. With the advantage of aftersight, can we believe that the citizens would willingly accept the entry of munitions ships into harbors where adjacent densely populated areas are particularly vulnerable? With afterthought it may be asked why such areas are exposed, why provision is not made for transshipment of great quantities of explosives, under which there are exposed relatively few persons who are not part of the working force. It is evident that the lesson of Halifax was not learned and applied.

The facts of the catastrophe are well known. The steamship *Fort Stikine* berthed at Victoria Dock loaded with a mixed cargo including about fourteen hundred tons of explosives. There was also baled cotton and a large amount of combustible fluids in drums, some of which were known to be leaking. Fire started in a hold from unknown causes; spontaneous combustion cannot be overruled; smoking discipline was imperfect.

Discovery of fire and the initiation of measures to control the blaze were tardy, and despite vigorous (but misdirected) efforts, the fires spread. Flooding measures were not resorted to; even if they had been taken, the depth of water in the dock would not have allowed for submersion adequate to wholly neutralize the explosives stored. The ship's engines had been shut down and under repair; thus, the ship could not move away. The ship's master was understandably loath to have the vessel towed from the dock and thereby lose benefit of the greater part of the fire-fighting force on the dock. Inevitably, the ship blew up. In two blasts, casualties were of the order of five hundred killed. Ten ships were destroyed or so badly injured that they were sold for scrap; altogether almost 35,000 tons of shipping were lost. Crew members, fire-fighters, and dock workers perished in large numbers, but principal losses occurred in the area adjacent to the dock where there were slums and low-income housing.

A report upon the incident dwelt upon the state of war permitting ships with explosives to be berthed unsafely. We think this is highly perceptive; a state of war relaxes regulations and allows a sense of fatalism (or false bravado in seeming to scorn danger). In the name of urgency, acts are performed and rules are breached to become costly lapses. It is probable that the *Fort Stikine* could have been berthed at a much safer position, although this would have entailed lighterage. The ship itself did not display tokens of its dangerous cargo; there was evidence that such failure was common, the masters arguing that it could identify to a foe the presence of a special cargo. But if the nature of *Fort Stikine's* cargo had been known to other ships' captains, might there not have been pressure to have *Fort Stikine* berthed elsewhere? Some evidence was brought out that signs of fire were first observed from a distant ship and that it was presumed by these observers that the *Fort Stikine* was aware of the fire. (This, before alarm was sounded on the *Fort Stikine*). If the nature of the *Fort Stikine* cargo had been known to these observers, it is certain more prompt and vigorous fire-fighting action would have been taken. The ship could have been towed to a less-exposed place. The court of inquiry pointed out improper stowage of cargo and the hazard of a shipment of several potentially dangerous materials. Cotton in a mixed cargo, especially when combustible liquids are included in the manifest, is a hazard in itself. Its inclusion with fourteen hundred tons of explosive reflected poor judgment. It is of lesser consequence how the fire started, whether by spontaneous combustion or stevedore carelessness. No evidence of sabotage was presented.

In our view, this is but an incomplete explanation. The cause goes back to wartime psychology that assures us that great things boldly done necessitate hazards. No sober thinking can suppose that the gains of expediting cargo movement outweighed losses. And we do hold that removal of the *Fort Stikine* from its Victoria Dock berth would have limited losses largely to that vessel. The historian will note the date as 1944, when the odds were fast mounting against the Axis Powers and the delivery of that particular cargo did not warrant desperate measures.

Failure at Outset by Authority to Appreciate Gravity

Other fires at these docks had been handled successfully. The fact that the *Fort Stikine* had a large cargo of munitions was not immediately recognized. In retrospect, we can say that the *Fort Stikine*, berthed illogically at Victoria Dock and not displaying hazardous-cargo signs, could yet have been towed to a safer distance. Any fire reported on such a ship should have called for immediate towing. The failure by any authority to take early effective action can be ascribed to absence of a central executive control with paramount power.

We are not sure which individual must take the blame for such absence of control, and as is common with many investigatory commissions, no one was charged. The compartmentation of jurisdiction, meant to relieve highest authority of onerous detail, did not provide means for crossing lines promptly. Yet the terrible loss of life and the damage to the war effort should weigh heavily on some consciences.

The stubborn pride of the ship's captain (he was lost) in not seeking to have the *Fort Stikine* towed and perhaps sunk is common to shipmasters. It is uncertain what responsibility the master bears for accepting and stowing cargo, but a harsh judgment says the captain was at fault; his alternative would have been to refuse command of the vessel.

The conclusion drawn from this disaster is that a conjunction of several dangerous conditions brought about the final catastrophe. The disaster was avertible and would never have occurred if good judgment had prevailed. The great bitterness of the whole lies in the noninvolvement of some who were accountable, and the sacrifice of a large number of persons who were in no way actors in the disaster.

10. Tragedy with Natural Gas

New London, Texas, School Explosion

On March 18, 1937, the large consolidated school at New London, Texas, was wrecked by an explosion, causing almost three hundred fatalities among teachers and pupils. Intensive investigations came to a reasonable agreement—a remarkable achievement—as to the circumstances.

New London is located in an area that has had great importance in petroleum production, the East Texas Field. Attendant casing-head gas was usually processed in gasoline recovery plants, leaving a "dry" residue gas roughly equivalent to natural gas. So plentiful was the supply of gas in this area that diversion was sometimes done by a consumer who might tap the utility's line with or without specific consent. In the case of the school's gas source, it is immaterial under what policy the fuel was obtained or whether competent mechanics made the connection from the utility's line to the school's appliances. This latter condition could be contributary but not controlling; the appliances were of an approved type. But the appliances (steam radiators) were not properly vented, so if the flame was accidentally extinguished and the fuel supply to the radiator's burner was not cut off, gas could accumulate in the vicinity of the burner.

The most incriminating evidence pointed to unventilated space where the gas line passed from the ground and under the school floor, and to a general condition of unvented hollow tile walls. Destruction by the blast made it impossible to say whether pipe connections were broken or faulty, or whether gas passed through an unlit burner. The locus of the ignition point was in an area poorly ventilated; ignition was attributed to sparking at an electrical switch in the manual-training room. The course of the explosion could be traced, and it pointed to the presence of a large amount of gas; the explosion was not localized, as would be the condition if a single mass of dynamite was detonated. It is possible that the diffused explosion was more destructive than a concentrated blast; it may have caused building collapse over a wider area, (many of the injuries were caused by falling debris). It was estimated that between five thousand and six thousand cubic feet of gas were involved. We may conservatively estimate that this amount (as an explosive mixture containing 8 to 10 percent hydrocarbon) would be the equivalent of half a ton of TNT.

In assessing blame for the disaster, it is hard to escape the economy-psychology involved in building the school. Authority that could distinguish savings in the use of gas-fired steam radiators instead of central heating would not be likely to obtain competent architectural guidance and supervision in the matter of avoiding hazards if gas-pipe connections were faulty. With the worldwide publicity given this avertible disaster, it should be impossible to have a recurrence, even on a small scale.

Natural gas was involved in another disaster, although there was no explosion. Liquefied natural was stored in tanks of the East Ohio Gas Company, Cleveland, Ohio, at minus 250° and 5 pounds pressure. On October 20, 1944, a tank failed and released a great volume* of the liquid. Ignition took place and flames enveloped industrial and residential buildings nearby. Most destruction and casualties (over 100 killed, 300 injured) were due to fire; explosions were subsidiary to the conflagration.

The lessons learned were the need for further study on the metallurgy of structure at very low temperatures and the need for isolating such storage tanks at a distance from inhabited

* Equivalent to thousands of times combustible as was involved in the New London school explosion.

buildings. The Texas school authorities and the Ohio utility both appear to have placed undue confidence in the mechanical integrity of their gas systems; neither gave adequate thought to the conditions that would follow the mechanical failure of such systems.

11. Munitions-Plant Explosions

The somber pictures of wartime munitions-plant explosions are similar. They depict large disasters, long casualty lists, and the recurring phrase "Cause not immediately known."* Whether the plants belong to friend, foe, or neutral makes little difference. The urgency of wartime conditions calls for a level of production of explosive material that is enormous in contrast to that of peacetime demand. Plants are enlarged; the number of personnel expands. The mere number of employees needed may site the plant near established housing, or housing springs up about the munitions plant. It is likely that the expanded personnel becomes increasingly vulnerable to mishaps. Regular peacetime employees are well aware of their hazardous occupation and are more likely to accept safety orders and measures. The supervisory force will probably have years of experience in recognizing potential hazards and in applying safety measures; wartime sees this exceptional management group diluted.

Of course costly peacetime accidents occur, but the lesser size of explosions may permit survival of employees, records, and equipment to give significant clues as to the cause of the explosion, a condition notably lacking in catastrophic wartime tragedies. The phrase above, "Cause not immediately known,"

*A sampling of disasters in such plants and munitions depots might read:

Dresden, Germany, December 1916
Burton-on-Trent, England, November 1944
Tessenderloo, Belgium, July 1942
Fort Smederovo, Jugoslavia, June 1941

is probably quite truthful, but in wartime it is considered injurious to morale to publish news of disasters; it may give aid and comfort to the enemy. A number of incidents serve to point up the inadequacy of precautionary measures and the possibility that lapses ended in disaster. Hard-earned lessons from nonmilitary industrial operations may not be disregarded but may be less than totally applied. We shall relate some, and in at least one can note the presence of a substance that, though believed quite safe, was destined to become a prime contributor to disaster.

a. The munitions plant at Morgan, New Jersey, was engaged in the production of molten amatol (a mixture of TNT and ammonium nitrate) and the loading of amatol into shell. Melting the two ingredients and some scrap amatol was done in steam-jacketed kettles at 194–203° F. The operation was considered safe.

On October 4, 1918, an explosion in a kettle initiated a fire in another area holding loaded 155 mm and 8-inch shell. An hour later there was mass detonation of about five hundred tons of stored shell. The dispersal caused other fires and explosions. Affected were storage facilities for five hundred tons of ammonium nitrate, which detonated. The human casualties were 64 killed, 50–100 injured. The most interesting feature of this catastrophe rests in the evidence that loaded shell can be detonated en masse, although previous experience showed it unlikely. Ammonium nitrate by itself had not been considered an explosive; yet when storage units for this chemical caught fire, the nitrate was detonated. It is not known whether a hot shell initiated the explosion. We shall see ammonium nitrate involved in other disasters. It does appear from research studies that by itself the nitrate is not easily detonated below its melting point. The presence of inflammable material seems always to be associated with explosions of ammonium compound.

The cause of the kettle explosion remains unsolved, but because it is certain the reaction started here, it is theorized that foreign substances (glass, nails, and so forth) were accidentally introduced and, by friction, during mixing created a spark (supervision of kettle loading was not strict). Friction and heat by other means was possible.

b. The US Navy maintained an ammunition and explosives storage depot at Lake Denmark, New Jersey. On July 10, 1926, lightning started a fire in a magazine holding 335 tons of TNT. Fire fighters responded promptly, but the unit soon exploded. Five minutes later, another magazine with eight hundred tons of TNT exploded. Great damage to other facilities was done by fire and blast. Nineteen persons were killed.

The investigatory commission found no evidence of human error nor of unduly sensitive explosive. It proposed greater spacing of storage areas; better lightning protection was presumed applicable. It may be pointed out that the petroleum industry has virtually eliminated the inherent danger from lightning striking storage tanks holding great amounts of readily inflammable fuel.

12. Nuclear Explosions

The ushering in of nuclear explosions is exemplified by the devastation of Hiroshima. It necessitated a new ordering of thought regarding man-made destructive power. It demanded a unit of measurement that was conveniently expressed as a *kiloton,* the equivalent of one thousand tons of TNT. Then followed the thermonuclear explosions, whose energy is expressed in *megatons*—millions of tons of TNT. Because the human mind has difficulty in correlating such forces with observed events, the thermonuclear explosion is passed by; it has no known beneficent application to human affairs. All detonations of this type have occurred at or near the earth's surface, and a large part of the blast effects are, in consequence, expended upward into the air.

As to nuclear explosions, the most astonishing (and perhaps happy) development has been in the direction of *diminishing force.* The most obvious expectation would be to see the size of nuclear devices increase until they covered the range of magnitude between the bomb dropped on Hiroshima and the thermonuclear device of Eniwetok. Could such devices ever be used except as weapons? The answer is "no." Turning in the other direction, research *has* successfully reduced the size of nuclear devices so that it is likely that a spectrum with effects ranging from a *strong conventional* explosive to that of a *mild* (Hiroshima-type) *atomic weapon* can be utilized in nonmilitary applications.

PLOWSHARE, the Atomic Energy Commission program to develop nuclear energy used in industry and science, has reported a series of tests that is worthy of study. The project includes in its many tests some nuclear blasts in the range of magnitude shown:

Test name	Equivalent Kilotons TNT
Sedan	100
Rulison	40
Gasbuggy	25
Handcar	12
Danny Boy	0.43
Sulky	0.09

Here is a sequence that is remarkable in that it encompasses explosions both greater than any man-made blast with conventional explosives and at the same time includes tests well within common applications of industrial explosives. Extension of the scale upwardly is, of course, demonstrable with military weapons; nuclear blasts below 1 kiloton are certainly practicable, but the economic choice between these and conventional blasting is not readily generalized.

Within the category of "beneficent explosions" are some nuclear blasts, actual or proposed, that are in effect engineering tasks done with intent to improve the condition of mankind. We believe it would be physically possible to dig a harbor, to excavate the route of a canal, or to change the course of a river by such explosions. It must be presumed that such events would be sanctioned only if there were no serious attendant hazard of fallout. There is no means of knowing the magnitude of individual explosions (a project such as a canal would require a series of blasts), but it is probable that the least of them would be comparable to a very large detonation of conventional explosive. We know of at least one attempt to use nuclear explosion underground in a body of rock. Objectives would include heating the rock to produce fluid hydrocarbons from solid carbonaceous matter, fracturing the strata to assist flow of fluids, and creation of a cavity either to function as a receptacle of fluid produced or to serve as a reservoir for fluid to be brought to the site and to be stored. We know of no such explosion made with proven economic advantage. The application is a clear challenge.

13. The Tunguska Meteorite

There is an age-old nursery story of Chicken Little, a barnyard fowl upon whom a wind-wafted feather chanced to alight; a near-panic ensued in the animal world because he thought the sky was falling. Some measure of this panic seized many humans on the morning of June 30, 1908, in the far reaches of Siberia northwest of Lake Baikal.

An object flashed across the skies with a brilliance rivaling the sun's and disappeared within seconds with sounds like massed artillery, which were heard at a distance of over six hundred miles. The engineer of a train on the nearest railroad (over three hundred miles distant) stopped his train, believing an explosion had taken place on it. A witness sixty miles south of the supposed place of fall felt heat from the flash. Earth tremors were recorded around the world. A vast cloud of smoke remained, creating atmospheric conditions that produced spectacular effects visible for many days. It became evident that the phenomenon such as a meteor entering earth's atmosphere had taken place.

Official investigation of the incident to determine the place of impact with earth awaited several years. This can be explained in part by the isolation (note distance from railroad) and because no casualties were reported. The region had great forested tracts, but the local economy depended heavily upon reindeer herds; many of these animals had been killed. Early expeditions entered the presumed area of impact with much difficulty. The scientists of the first and later expeditions discovered a region of about one hundred square miles where the forest was more or less damaged. In some places, the forest was flattened, sometimes burned; in others, the trees were denuded of all foliage. Large craters were absent; meteoric remnants were not recovered.

The tentative conclusion was that a meteor had entered the earth's atmosphere at a low angle and was consumed in flight. The explosive damage to the forest was due to the head wave of the object. The destructive effect of the meteorite may be ascribed, in large part, to the air compressed by and pushed ahead by the object. The resulting airwave becomes the equivalent of an explosive release. It was impossible to determine accurately the size of the meteor, but it was estimated that the object had a mass of a million tons and entered the atmosphere with a velocity of twelve miles a second (March 50). It has been calculated that the energy involved was equivalent to that which destroyed the island of Krakatoa. Other calculations placed the disintegration of the Tunguska meteorite as equivalent to the energy of two to thirty megatons of TNT. It has been postulated also that the phenomenon was that of a comet entering our atmosphere.* This is consistent with the absence of cratering.

The Tunguska meteorite (if such) is indeed a remarkable example of an explosion. The absence of loss of human life points this up; even economic loss was minor. However horrific the mental picture of the meteor falling in a metropolitan area, Chicken Little can take comfort from the odds of one in a million that such a meteor would find such a place. Another estimate has it that United States residents may expect one fatality each ninety-three hundred years. Tunguska is unique in our history; this is our only encounter with extraterrestrial objects of any considerable size.

*A comet is presumed to be a dense cloud of small particles, i.e., mineral matter and frozen gases.

14. Fertilizer, a Benefactor to Agriculture, Demands a Price

It is well known that three prime requisites of commercial fertilizers are compounds of nitrogen, phosphorus, and potassium used in ratios best suited to particular crops. We shall be concerned here only with nitrogen-containing chemicals. The most concentrated form of nitrogen compound is ammonia gas (NH_3), which is over 80 percent nitrogen. Its water solution may be used as a fertilizer. For convenience in storage and handling, the ammonia may be converted chemically into stable solids such as ammonium nitrate and ammonium sulfate, containing 35 percent and 22 percent nitrogen, respectively. The ammonium nitrate is well known as an explosive, but when associated with the sulfate, its stability is increased and its explosiveness is greatly diminished.

The Badische Analin und Sodafabrik (BASF) at Oppau, Germany, produced great quantities of the double salt ($[NH_4]_2SO_4 \cdot NH_4NO_3$). On September 21, 1921, about 4500 tons in storage exploded. The casualties were 565 dead, 1500 injured. The material damage exceeded 30 million dollars.

The double salt was stored as a solid mass; the chemical stability was so assured that it was customary to use mild explosive to break the mass for subsequent handling and shipping, much as might be done in this country for sulfur, which is poured as molten sulfur into storage forms where it congeals to a great solid mass. We state that the stability of the double salt was assured; the standard tests for explodability were considered quite severe and had demonstrated safety. The cause of the explosion is unknown; conjecture remains conjecture. Perhaps there was an inadvertent increase in the ratio of nitrate to sulfate, which might have made the mass more sensitive to the blasting, or the accidental presence of organic matter, which might have been consumed violently by the nitrate and produced enough heat to favor the explosive reaction of the nitrate in the main mass of double salt. We can never know whether human error caused the catastrophe. The terrible toll exacted gives fair warning that the nitrate must ever be treated as an explosive, regardless of its name and intended purpose. We shall see that disaster has followed neglect of this lesson.

A. A Lesson Ignored?

On April 16, 1947, two freighters taking on cargo at Texas City, Texas, exploded and exacted a toll of 570 people killed, 3500 injured, and property damage upward of $100 million.

Texas City has a complex of chemical and petroleum industries and is a location well suited to the transportation needs of these industries. The French freighter *Grandcamp* was taking on, amongst other cargo, a load of fertilizer in the form of granular ammonium nitrate (NH_4NO_3). About 2200 tons of the fertilizer had been loaded when fire was reported in a hold. After twenty minutes of ineffective fire fighting by the immediate personnel, it was decided that the fire(s) would be smothered with steam. Such a measure to prevent access of air to the combustible had the evident advantage that it (steam) was likely to cause less damage to cargo than a stream of water.

The fire appeared to worsen, and ten minutes after the steam was first applied, the Texas City fire department was summoned to help. Thirty minutes later, the *Grandcamp* blew up. The immediate destruction was immense, damage being increased by fire spread by the dispersed ignited cargo of the *Grandcamp*. One fire, in particular, had been set on the freighter *High Flyer*, which also carried a part cargo (1960 tons) of fertilizer. Fire in the *High Flyer's* hold was reported about nine hours after the *Grandcamp* exploded; seven hours later, the *High Flyer* blew up, killing many rescue workers summoned to the *Grandcamp's* disaster.

Many of the casualties were in adjacent plants and properties, many in the town itself (population 17,500), and many amongst spectators drawn by the *Grandcamp's* explosion and others in the vicinity of the *High Flyer*. Such were the background events.

In retrospect, the application of steam to smother the fire was a mistake. It tended to raise the temperature of the hold's cargo to a critical point. Unanswered is the question of how the fire started. Perhaps a minor blaze from prohibited smoking, or possibly the stowage of sacked fertilizer against or in too close proximity to a steam line. Fighting of fires with smothering steam is often very effective in petroleum operations (and, as noted, Texas City is a petroleum-processing area, so dockside personnel may have drawn on this experience).

There is no proof of instability of the NH_4NO_3, made by Monsanto Chemical Company. Yet the failure to recognize the potential hazard is astonishing in a technical minded organization. It is easier to understand the indifference of crew members toward cargo.

We may assume that there was great disruption on the *High Flyer* after the *Grandcamp* exploded, yet there was undue delay in measures to neutralize damage. Efforts to tow the *High Flyer* were futile, but there was an interval of about five hours between the report of fire on the *High Flyer* and explosion of the *High Flyer*. Was it possible to flood the hold? Was depth of water adequate to submerge the hull?

We are less charitable toward the *Grandcamp*. First report of a fire was given at 8:10 A.M. Smother steam was applied at 8:30 A.M. A twenty-minute interval of exploratory and (apparently) ineffective fire fighting against a severe latent hazard indicates inadequate organization.

This is not to say that those in charge should have regarded the nitrate as an explosive *per se,* but that they should have known it could play a part as a *component* of an explosive mixture. (Note the parallel role of *potassium nitrate* in gunpowder.) In other words, the fertilizer, NH_4NO_3, in the presence of a burning combustible, would greatly increase the intensity of the combustion process. The *Grandcamp* carried solid combustible cargo; once its wrapping was ignited by any means, combustion would be sustained by access to the fertilizer.

B. Conclusion

Monsanto possessed enough background information to have made all concerned aware of the hazards. City, state, and federal authorities bear an indeterminate share of blame. The failure to neutralize the *High Flyer* compounded disaster. It is our opinion that fire in other combustible cargo came in contact with ammonium nitrate. Once ignition between the two had begun, it would be self-sustaining.

Preventive measures: recognition of this fertilizer as an explosive medium, proper stowage; fire discipline and control; acceptance of immediate towing, flooding, and so forth by any involved vessel.

15. Dynamite Disasters

Cali, Colombia, was the scene of a terrible catastrophe August 7, 1956, when an explosion of dynamite under transport through the city killed over four hundred persons and injured a thousand more. About thirty-five tons of industrial dynamite were being carried on seven military trucks.

The disaster was said by some to be the result of politically motivated sabotage. We agree that the political atmosphere was such that sabotage cannot be excluded as a possibility, but even were sabotage acceptable as an immediate cause, the responsibility goes deeper. The explosion of commercial dynamite without an accompanying detonator is unlikely. If one of the party to the sabotage was with the trucks, he must have perished. How shall we condone the transport of explosive in peacetime through a congested area by a *convoy* of trucks? Is this military arrogance or ignorance? Surely, if explosives must be moved through a city, its transportation should be regulated by civil authorities and vehicles carrying explosives should travel at a reasonable distance from one another so that the explosion of one unit will not cause sympathetic detonation of the remainder. Cali had a population of 250,000; the locus of the explosion was a poorer section of the city. The accident occurred at night; the destruction of utilities and the absence of adequate rescue and relief facilities added greatly to the suffering. In retrospect, some of the lessons from the relatively minor tragedy of the trolley car incident were applicable.

Santander, Spain, saw a major catastrophe compounded of

illegality, official lapses, and spectator curiosity. The ship *Cabo Machichaco* arrived at Santander, Spain, on November 3, 1893, with a mixed cargo including over thirty-two tons of dynamite (undeclared). The vessel was moored at the quay in violation of regulations governing ships carrying explosive. Fire had broken out on the *Cabo Machichaco,* and a great crowd of people swarmed on the quay to watch. A small amount of dynamite was unloaded, the agent stating that removal was complete. Soon after, the vessel blew up a cost of about five hundred dead and two thousand injured.

The disasters at Halifax, Texas City, and Santander display a common theme in counting casualties: apparently civil authority had done nothing to curb sightseers' entry into a hazardous area.

16. Novas

Thermonuclear release of energy has long passed our comprehension of magnitude, and we take refuge in stating it in terms of megatons of TNT. But if most of us have never seen an explosion of a ton of dynamite, of what significance is an event of a million tons? We must accept some natural events with philosophical awe, if not understanding. T.S. Eliot would have it:

> This is the way the world ends
> Not with a bang but a whimper.

Be it so; but in its place are offered events occurring and recurring—the phenomena of novas; new stars. The release of energy by them is calculable only theoretically. Consider a mass of TNT compressed into a single bomb of earth's weight; its explosion in a nova would pass unnoticed.

17. Conclusions

The thousands of untoward "accidents" are but small arrows pointing to the conditions favorable to the few catastrophic events. It is logical that we should apply our hard-earned lessons gained from explosions of man's making and of those taking place in his artifacts.

Should not the genius that has discovered how to obtain energy and to use it beneficially be equally capable of controlling one direction of this energy and of preventing its disastrous release?

It is not inferred that natural events are controllable by man, yet some measure might be obtained. For example, nonexplosive eruption of volcanoes accompanied by great flow of potentially threatening lava might see amelioration by the diversion of the flow. This has been done to a small degree. Perhaps in the future man will be able to recognize the conditions of impending cataclysms akin to Krakatoa and Coseguina and be able to lessen the magnitude of disaster.

We close with this belief: Explosions in our mundane affairs are controllable and avertible; the choice of action is ours.